Game Theory

A Beginner's Guide to Strategy and Decision-Making

John Ledlin

Table of Contents

Introduction 1

Chapter 1: What Is Game Theory? 4

Chapter 2: The Applications of Game Theory 14

Chapter 3: Prisoner's Dilemma 28

Chapter 4: The Shapley Value 35

Chapter 5: Battle of the Sexes 42

Chapter 6: The Centipede Game 47

Conclusion 55

Introduction

It seems odd to turn to mathematics for help in social interactions. Mathematical principles do not appear applicable to social situations, as interactions and relationships are governed by or depend on emotions. However, game theorists would argue differently. It is through understanding our emotions and being rational about our feelings in social contexts that forms the basis of game theory.

For example, how do you cut a cake so that everyone feels satisfied with their piece? It is not as straightforward as we like to think. It was only in the 1950s that George Gamow and Marvin Stern formalized the logical technique to cut a cake for two people. This is known as the *divide-and-choose solution*. Person A cuts the cake. Person B chooses first. This forces Person A to be fair and divide the cake equally. Moreover, Person B is satisfied, as they get to choose first.

A decade later, John Selfridge and John Conway devised a method for cutting a cake for three people. Person A cuts, and Person B trims, as it is unlikely that the original cut will result in an exact third. Then, Person C picks first. Next, Person B selects their piece, and Person A takes the remaining piece. Moving onto the trimmings, Person C now cuts first, while Person B selects their trimming piece first. Then Person A chooses next, and finally, Person C gets the remaining piece. Since the cutter never chooses first, it forces them to divide the cake fairly. What is interesting is that no reasonable or rational method has been discovered to divide a cake for larger groups, such as for weddings, birthdays, or anniversary celebrations. Therefore, we are currently doomed to not have equal cake slices at these celebrations. The injustice is unavoidable.

The envy-free cake-cutting concept may seem unimportant. We may not care about how much cake we get at a party. However, we are thinking too superficially. After all, the cake simply represents a resource. Certainly, a cake is a trivial resource. However, when it comes to water, time, money, and energy, then suddenly it becomes necessary to learn how to divide a cake in a way that satisfies everyone.

The envy-free cake-cutting principle is just one of the many diverse game theory principles. It considers one specific social interaction. However, there are numerous theories which exist in game theory, such as the prisoner's dilemma, the Shapley value, Arrow's impossibility theorem, and Deal or No Deal. Each of them deals with a specific social context. The resolutions for these principles differ immensely from one another, as each social interaction or context is unique. As a result, each social context or interaction has a different goal. Thus, game theory is an extremely useful branch of mathematics, as it first teaches individuals the objective of each social interaction. Next, it reveals the logical resolutions which bring about the attainment of this objective.

Though game theory was only officially formalized in the last decade, it has since become a popular branch among thinkers, researchers, and even psychologists. In the last two decades, numerous books have been written on the subject responding to the demand of this field. The demand is growing as individuals are beginning to recognize that managing both smaller and larger social interactions helps to create a better world, promote fairness, and allows them to lead happier lives with stronger relationships. While game theory is based on understanding and analyzing social contexts or interactions, it advises against manipulation.

Game Theory: A Beginner's Guide to Strategy and Decision-Making aims to explain the concept of game theory and why learning about this branch of mathematics is immensely useful and beneficial for everyone. It will cover the main principles, such as the prisoner's dilemma and Shapley value. Finally, it will explore how game theory can be applied to both smaller and larger social contexts. As mentioned, this subject does not have the goal of encouraging manipulation, but rather managing social contexts to promote fairness, cooperation, or necessary competition.

Chapter 1: What Is Game Theory?

Basic Definition

A brief definition of game theory is an analysis of social interaction models to aid decision-making for these social contexts. It then applies logic to predict decision-making for the given social context to produce the most desirable or advantageous outcomes.

Yet, as each social interaction or context is unique, these situations or contexts have different goals or objectives. Therefore, as the objective for each context is unique, there are different strategies applied to produce the best outcomes. In short, one technique cannot simply universally be implemented. Each social context demands thorough investigation, and varying methods will be applied.

As game theory employs logic and rational thinking to arrive at the best outcomes for each social context, this subject has been considered a branch of mathematics. However, as it also deals with social interactions both on a small and large scale, it is viewed as a subject group of social sciences. Some of the principles discuss employee remuneration and benefits, making it also a subcategory of economics and social economics. Game theory, thus, is a type of thinking that incorporates various disciplines. It is also sometimes seen as a branch of psychology, as it explores the best outcomes related to a social interaction. In other words, determining which outcome promotes the best psychological wellbeing of the individual and individuals involved is a feature of game theory.

History

The origins of game theory are quite difficult to pinpoint. Some historians claim that its origins can be traced back to the late 16th century. Gerolamo Cardano, an Italian polymath, is credited with writing about the dynamics of games in 1564. As Cardano used gambling to sustain himself, he became interested in the role luck plays in games. Furthermore, he tried to devise effective ways of cheating to aid him in this goal. For the next few centuries, several mathematicians such as Blaise Pascal and Ernst Zermelo devoted themselves to studying games critically. Pascal specifically concerned himself with the role chance plays in games, while Zermelo devoted his investigations to the workings of chess.

Twenty years after Zermelo's work on the dynamics of chess, a new type of thinking appeared in game theory, namely the concept of the zero-sum game. While it is commonplace now to hear the term, it only entered the Oxford Dictionary in 1944. John von Neumann was the mathematician who was credited with using this term for the first time. Von Neumann was an Hungarian-American mathematician and physicist who devoted his life to mathematical models and pure mathematics. To this day, he is considered one of the finest mathematicians of all time. It was his investigation of the zero-sum game that coincided with the formal establishment of game theory. Von Neumann's work involved investigating zero-sum games and the best strategies individuals should apply in such interactions to produce the best results for themselves. Nevertheless, with the creation of the zero-sum-game theory, the theories of different types of games were also established. For example, the win–win, no deal, lose–lose, and win–lose theories. Win–lose is another name for the zero-sum game. Von Neumann's later work then shifted from zero-sum games to win–win situations. In the next section, we will look at the models or bases of different games.

Following von Neumann's contributions, mathematicians Merrill M. Flood and Melvin Dreshner revisited von Neumann's work on the zero-sum game and came up with the prisoner's dilemma theory in 1950. The prisoner's dilemma is one of the cornerstones of game theory, as it describes how the zero-sum-game model is applied to a specific context in society, namely the choice a prisoner must make. Like with most game theory solutions, they are complicated and specific to the situation. The most notable solution to the prisoner's dilemma is the Nash equilibrium, named after the mathematician John Forbes Nash.

For the second half of the 20th century, with his careful attention to zero-sum-game models, Nash contributed immensely to the subject of game theory. In 1994, he won the Nobel Prize in Economic Sciences. Both von Neumann and Nash have been seen as key figures in the creation of game theory principles.

Meanwhile, Lloyd Shapley was producing his work on strategies and decision-making for cooperative games. As a result, another key principle of game theory, the Shapley value, was devised thanks to Shapley's contribution. Later in 2012, Lloyd Shapely and Alvin E. Roth, an economics professor at Harvard and Stanford, earned themselves the Nobel Prize in Economics for their work.

In summary, game theory can be said to have been initiated or formalized in the 1940s. This makes it a young and newly established branch of mathematics, economics, and social sciences. It is also a continuously growing field as various economists, mathematicians, and thinkers are revisiting basic game theory strategies. One example is the living economist Thomas Sowell, who compares how zero-sum games and win–win situations affect social relations and demographics. Thus, while game theory is a relatively new field, it has exploded in the last century. This coincides with the formal recognition of economics as a field of study, as it was only in 1968 that the Nobel Prize in Economics was added to the original five Nobel Prizes.

Different Types of Games

What Is a Game?

A game is composed of several key characteristics. First, it involves players. These are individuals who are involved in the events, and their decision-making influences the outcomes. A game is dependent on two or more agents who are involved in the decision-making. If there is only one person who decides, this is not a game, as there is no other agent responsible for determining the outcomes.

Next, there are general strategies or strategies per player. Each of the individuals is given a certain amount of choice or freedom to determine the results of the social interaction. In addition—and not to be confused with the strategies—there are the pure strategies of Nash equilibria. Nash equilibrium refers to the methods that, if implemented, allow an individual to achieve the best possible results of the particular game (social interaction).

Finally, the last characteristic of a game is outcomes. These are the events which are produced following all the strategies the involved players make. It is a byproduct of the interdependence of players' choices. An outcome typically embodies the reward or loss which the players experience. From the very beginning, players bear in mind the loss or reward they are to reap after employing a chosen method or methods.

On top of this, there are some other characteristics which occur in some games, but not in others. The role of luck or chance is a factor in some games, but not all. Consider chess. Chess is a game which does not involve luck or chance in any way. Even if one of the players makes a costly decision, this decision is considered a strategy implemented by the agent and not dependent on the role of chance or luck. This differs to card games such as poker which are subjected to chance. The players are dealt cards randomly. In chess, there are a set number of pieces—which are restricted to specific moves—thereby eliminating the role of chance or randomness.

Moreover, some games are characterized by whether players have access to the strategies of the other player or players. If they do, this game is said to have complete information. If not, it is referred to as having incomplete information. Chess is an example of a game which involves complete information. As the pieces, a rook or a knight, are constrained to a specific set of moves, the players can learn which moves are available to the positions of their opponent's pieces on the board.

Lastly, the word *game* has been used quite loosely here. A game can refer to any situation in which several participants are involved in employing different strategies to achieve a specific goal or bring about desired outcomes. Based on this definition, a game does not deal specifically with a mutually agreed-upon scenario where individuals work to attain a goal for entertainment purposes. It describes any situation in which the people rely on decision-making to obtain that result. Therefore, you may be playing a game and not be aware of this fact.

Nash Equilibrium

Throughout this book, we will look at the Nash equilibrium of all the various social interactions in game theory, and the Nash equilibria that are the best proven outcomes for each player in the context. Therefore, these are strategies that are said to bring both players the most advantage from the interaction. Typically, the strategies are the same. Mathematicians have devised methods which both players should follow which will produce the best outcomes. Yet, they are not universal. Therefore, the Nash equilibrium is specific to each context and should not be applied to other interactions.

As stated, mathematicians have over time developed formulae to make their cases. However, it should be acknowledged that these proven Nash equilibria are not always perceived as the morally correct option. Like in many cases of the Nash equilibria, they are often in favor of one's own interest as opposed to mutual benefit.

Game Types

In this section, we will look at the basic models of games. Whether it is business, foreign policy, or entertainment with friends, if there are two or more parties involved in the results, it will follow one of the below models.

Cooperative and Non-Cooperative Games
A cooperative game is one that forces the players to negotiate, make a deal, or reach a consensus to achieve desirable outcomes. The attainment of the best outcomes is contingent on the two individuals being able to reach an agreement. Team-building events best exemplify cooperative games. A successful outcome can only follow if the team members can negotiate between themselves and take on roles to effectively manage a successful result. Furthermore, a business relies on cooperation between employees to obtain the most desirable results: an increase of

profit, customer satisfaction, and efficient production or productivity. This is why team-building projects are undertaken in companies, as they want to improve workers' skills in cooperative games.

A non-cooperative game is one that involves competition. Cooperation, either through circumstance or choice that the parties have made, is not possible, and the two agents or players have to follow and implement strategies that produce the best outcomes for themselves. The prisoner's dilemma is an example of a non-cooperative game. Cooperation or negotiation between the prisoners is not possible, as they are held in separate cells, and they cannot communicate with one another. Therefore, both are forced to apply strategies which produce the best results for themselves, even if it brings about negative results for the other prisoner. In the third chapter, we will discuss the prisoner's dilemma more extensively.

Constant-Sum, Zero-Sum, and Non-Zero-Sum Games

Zero-sum and non-zero-sum games bear some resemblance to cooperative and non-cooperative games. However, the main difference is that constant-sum, zero-sum, and non-zero-sum games focus on the outcomes, while cooperative and competitive games also consider the circumstances. Earlier, we looked at the prisoner's dilemma. It was the environment or circumstances that made cooperation impossible.

A constant-sum game is one in which the outcomes remain constant. A zero-sum game is an example of a constant-sum game, as the outcomes do not change.

A zero-sum game is a game in which, if one party or player wins, the other party must lose. There cannot be two winners. It is called a zero-sum game because, if you add the outcomes of the one person who has won and the outcomes of the other person who has lost, the result is always zero. Thus, it is a constant-sum game, as it always produces zero. Examples of zero-sum games are poker. The objective of the game is for one agent to walk away with all the other players' chips. It is still a constant-sum game, as the amount in the pool remains the same for whoever wins.

A non-zero-sum game is the opposite. Another name for this category of games is win–win. The gains of the one player do not bring about a loss for the other player. Thus, when you add their gains together, it does not equal zero, which is why it is known as a non-zero-sum game. Trade is an example of a non-zero-sum game, or win–win. For example, Country A opens its markets. It can trade with countries throughout the globe for the latest developments in healthcare, technology, and agriculture. Furthermore, Country B, which trades with Country A, also gains as Country A buys their goods. Thus, both countries walk away with more than they had, and the results do not equal zero.

Symmetric and Asymmetric Games

Symmetric games are those in which the same strategies are available to both of the players. A symmetric game generally is a normal form or short-term game, as there is only one round or two rounds of decisions. If there are too many rounds of decision-making, generally the games become asymmetric because dissimilar strategies appear. An example is candidates applying for a job interview. All the candidates have to follow the same procedure or processes to apply for the job. They have to complete an application form and write a cover letter.

Asymmetric games do not involve the same decision-making for both players. Furthermore, the decisions open to both players may not bring about the same results. Entering the market and gaining market share between different companies is seen as an asymmetric game. For example, Walmart currently has the biggest number of employees and differs from Amazon in terms of their marketing strategies and production. Amazon can rely more on algorithms to maintain customer satisfaction, whereas Walmart has to rely on adequate training of human resources to bring about customer satisfaction.

Normal-Form and Extensive-Form Games

Earlier, I spoke about some characteristics which appear in some games and not in others. I purposefully did not bring up the topic of time. Time is a crucial element in games.

A normal-form game is one that is not influenced by time. It is generally represented as a matrix. On the x-axis, the options or strategies and outcomes of Person A are displayed. The y-axis fills in the methods available and possible results for Person B. If there are two possible strategies for Person A and Person B each, there will be a total of four quads making up the matrix. Each quad represents a different strategy and the outcomes it produces. Using this matrix, the best strategy is chosen for both Person A and B.

An extensive-form game is one which is affected by time. This game is represented on a tree-like diagram. Each branch or node of the tree represents a different decision which produces a secondary series of decisions. If a specific decision is made, it will create further possibilities of choices which both parties or players will have to make.

Summary

Game theory has seen a unique and recent evolution. As mentioned, it was first informally created by Cardano, who wished to maximize his gambling abilities. Since von Neumann,

Nash, and Shapley, studying games has become more established. The four basic varieties of games, which were discussed in the previous sections, outline the structures of different game types.

In the next section, we will look at how these basic structures differ in various games, making them a unique type of social interaction. In fact, it is the changes to these constants (strategies and outcomes) which make each game particular and interesting. We will then also look at examples of Game Theory situations and how mathematicians resolve these challenges or play these games.

Chapter 2: The Applications of Game Theory

Introduction

At a primordial level, games hold much value for us. There are a variety of games available in society, including traditional board games such as Monopoly, smartphone sensations like Crazy Birds, Candy Crush, or Pokémon Go, and even the biggest sporting events such as the NBA or English Premier League, which sees viewership into the millions or billions.

Traditional board games such as Monopoly have been adapted, like the latest version of Monopoly Deal, or like Risk, which has been redesigned to a Game of Thrones theme, adapting to popular trends in entertainment. As mentioned earlier, even team-building events which have been implemented recently in enterprises throughout the world involve a game. Therefore, at every level of entertainment and on some levels of necessity, such as team building, one finds games. Play is an expansive and multi-faceted industry.

However, if you strip away the Game of Thrones appearance of the modern-day Risk adaptation or look beyond the colors of your national sporting team, you will see that the structures of these games or events bears much similarity to the four models of games we spoke about in the last chapter. While the previous chapter attempted to answer the question "what is a game?" this chapter will try to answer the question "why do we play games?"

The Psychology of Games

Since gaming or playing is a phenomenon that occurs in every continent and even across different mammals—typically social creatures—one must consider this activity as more than something which creates fun. Rather, it provokes the question "why are games fun?" The simple answer is that games allow us to play. This simplistic answer forces two further questions: "what is play?" and "why is play fun?"

Peter Gray provides an explanation of play, addressing the two questions:

> "Play is a concept that fills our minds with contradictions when we try to think deeply about it. It is serious, yet not serious; trivial, yet profound; and imaginative and spontaneous, yet bound by rules. Play is not real, it takes place in a fantasy world; yet it is about the real world and helps children cope with that world."

There are two key aspects which Gray mentioned in the explanation above: "bound by rules" and "it is about the real world."

Rules

All games are bound by rules. If you buy a board game, you read a printed copy of rules which help you to learn to play the game and how to play it effectively. Games such as Monopoly and Clue have long existed in tradition so that most people are familiar with the rules.

Card games also have rules. For example, Poker has an established rank of hands which grade what the best five cards for the table are. A typical rule that seems to exist across many games is turn-taking. This is the case for card games, board games, and others like chess, darts, and snooker. Therefore, rules compose an essential element of games.

While it is true that different games have different sets of rules, what remains constant is the existence of rules. It is true that rules are crucial to ensure fair play; however, the function of rules goes beyond simply promoting fairness.

The rules define the game. They shape or give a structure to the game, making it unique. Consider soccer and basketball. Fundamentally, what distinguishes soccer from basketball is that the players must use their feet to control the ball (with the exception of the goalkeeper) and the latter requires the players to use their hands to control the ball. If soccer players started using their hands instead of their feet, the divisions between basketball and soccer would become quite blurry. Consider yet another example. If we started using a bow and arrow to aim for a dart board, instead of using our arms to guide a dart onto a specific position on a dart board, there would be little to define archery and darts. Archery uses bows and arrows and targets. Darts uses darts and targets and no other devices. Thus, like these two examples demonstrate, the rules make the game. Without them, these games could not be played and would not exist.

As stated above, rules also promote fairness. There is a deep psychological need for fairness. A sense of justice exhibits in not just humans, but also our evolutionary cousins, chimpanzees. Jonathan Haidt, an American social psychologist and professor of ethical leadership at a New York university, explains in his TEDx lecture that the concept of justice is ingrained into humans. We, at a very primordial level, want justice to be present in society.

By this logic, fairness is one of the goals games try to achieve with the creation of their rules. Moreover, it could be argued that games function as a kind of practice ground for concepts such as justice. From a young age, we teach ourselves and young children how to play by the rules so that they learn about concepts of fairness and justice. Not only do they learn it, but they become part of the process of instilling these rules in society. For example, if they sense that their fellow player is not following the rules, they call them out. They express the injustice with statements such as "you're cheating" or "that's not fair."

Representing Real Life

It is not only games which are bound by rules, but also real life. In real life, they are called laws, policies, or regulations. For instance, whether we own a Ferrari or a station wagon, there are restrictions on what speed we must travel at as we are sharing public roads with other motorists. Most nine-to-five jobs are characterized by the rule that employees have to start working at nine and end their day at five. Finally, you are permitted to travel to most countries in the world, but you are required to follow flying regulations. You cannot smoke on board, typically you need to turn your devices onto flight mode, and you are not permitted to pack flammable or aerosol items.

The reason we play games is that they teach from a young age about boundaries and limitations. It is preparing us for the real world, where, to participate in smaller and larger social interactions, we need to be restricted. For example, Clue is a detective game that reflects crime and investigation. If you guess "who-done-it" correctly, you win. If you guess incorrectly, you lose. This very much mirrors the social need to have police members and investigators put the right person behind bars.

First, if we do not comply with the rules, we cannot play the games. In this way, games teach us from a young age how to follow rules. They also reward good behavior. If we follow the rules and we apply the rules in a way that is to our advantage, then we are generally rewarded. In computer games such as Tomb Raider, the player has to solve a puzzle to move to the next level. Tomb Raider rewards problem-solving and the ability to solve puzzles. Monopoly teaches children that property is a source of income. Hotels on properties bring in the most income. It helps players to learn how to spend their money wisely and to develop a portfolio of assets.

Social Skills

Going by the definition from the last chapter that a game is an activity which involves two or more players who employ strategies to produce specific outcomes, this definition limits games to only social interactions. Using this explanation, games are playgrounds or test grounds for individuals to better their social skills. First, they teach us how to cooperate. For example, role-playing games like Dungeons and Dragons rely on a team of heroes or mythical creatures such as elves and orcs to complete a quest through cooperation. The same is true for Dota. As one of the most popular internet games, Dota requires teams to choose different heroes to beat the other team. Each player who controls one of the heroes has to learn which hero works best for their team and needs to learn how to work with the team to reach the objective.

On the other hand, there are some games which teach the skill of analyzing or trying to understand people. These are called social deduction games. Poker, specific Murder Mystery games, and Secret Hitler are examples of this subdivision. Poker is not a pure social deduction game, but it certainly involves reading people. Pure social deduction games like Secret Hitler get people to analyze their friends' behavior to discover who is Hitler, who are the fascists, and who are the liberals.

This is a skill which is necessary in real life, as we need to try to read people to learn if they can be trusted or whether they will do us harm.

All in all, both games and play are intrinsically valuable, as they get us to learn the dynamics of cooperation and reading people. They also teach us how to balance cooperation with trust. If someone does not play fairly, this often transcends the game and could be a clue as to how they will treat us in society. Therefore, games reflect many characteristics of real life. They aid our development and socialization. When joining society, we know it is necessary to cooperate—and thanks to games, we have acquired some basic cooperation skills—and how to read or analyze those we encounter.

The Games We Play

In the last chapter, we looked at four basic features of a game. Whether it is a strategy, arcade, social deduction game, or a sport, it is characterized by the presence or absence of these features.

In this section, we will consider how these features distinguish popular games and explain how they apply to social contexts of everyday life.

Zero-Sum and Non-Zero-Sum Games

Games can either result in a zero-sum or a non-zero-sum result. For instance, Rock, Paper, Scissors is a zero-sum game. There can only be one winner. Poker is a perfect illustration of a zero-sum game. A more basic version of poker, known as Kuhn poker, is used in game theory. In Kuhn poker, the deck only contains the three picture cards (a jack, queen, and king). Only one card is dealt to the players. The players go through a round of betting. Once the betting is concluded, the player with the card of the highest value wins the round and all the total of all the bets placed during the round. For every round, the pot goes to the player with the highest cards. The objective of the game is to get all the money from the players into one pool—played in rounds—into the hands of one person. A person walks away with the winnings. Thus, the aim of the game is to engage in competition and to try to obtain all the resources. It is a zero-sum game, as the amount acquired by the winner when added to the loss incurred by the other players equals zero.

The traveler's dilemma is an example of a non-zero-sum game. In this interaction, there are two passengers flying with identical-looking suitcases and which contain the same goods. In the instance of the airline losing the two suitcases, the airline offers the passengers insurance. The capped amount that the airline will pay is $100. However, the airline intends to reimburse the passengers according to the exact value of the contents of their luggage. Subsequently, the airline manager asks both passengers separately for the total price of their luggage. As the strategies of each traveler remain hidden from the other player, the two players cannot conspire to bring about a mutually beneficial result. If Passenger A and Passenger B both write down $100, then they receive $100.

Yet, the airline has implemented a clause that, if both passengers write down a different figure, they will pay out according to the lower figure. On top of this, the airline will include a penalty of $2 for the passenger that wrote the higher figure and include a reward $2 for the passenger who wrote a lower figure. This is the airline's method of incentivizing honest behavior and punishing travelers who seek to make a profit from a dishonest claim. In this scenario, if Passenger A claims $100 and Passenger B $99, Passenger A will be penalized $2, resulting in them receiving $98. Passenger B, who wrote down $99, will get $101. Thus, the inclusion of a penalty and reward make things complicated. Both travelers will attempt to outthink the other, writing down lower and lower figures so that they optimize their outcomes. It may be hard to believe, but the Nash equilibrium, or the figure which is said to result in the optimum outcomes, is $2. Though economists still maintain that going for a higher figure is a better result, the Nash equilibrium first reveals that the two travelers will keep trying to outthink the other. Furthermore, in the case of whether the penalty and rewards are much more than $2 and $50 instead, the individual who writes down $2 seeks to gain $52. The passenger who writes down $50, in this case, will receive the $2, as it is a lower amount, and penalized $50, making a loss.

It is true that the traveler's dilemma is a unique set of circumstances. However, situations like this play out, for example, every time Harry Potter books were released at the height of their popularity. The demand for them was huge. People queued outside of bookstores to make sure they would get their copy as soon as it was released . However, individuals aimed to avoid waiting in the queue and wanted to be first in line. If the bookstore opened at 9am, they would aim to be there at 8:59. Another potential buyer would try to outthink the others and arrive outside the bookshop at 8:58. The pattern of trying to outperform other buyers resembles very much the dilemma the two travelers face.

Cooperative and Non-Cooperative Games

Cooperative and non-cooperative games often closely resemble non-zero-sum and zero-sum games. However, there is a significant distinction in the outcomes. A zero-sum game always ends in a result of zero. If one player gains, the other must lose. A non-zero-sum does not produce a result of zero. Take the traveler's dilemma, for instance. Even if Traveler A writes down $2 and Traveler B $3, Traveler A will receive—including their reward—$52 and Traveler B—including their penalty—$47. The sum of the two results is +5. Thus, the result is not zero. It is not a zero-sum game. And also, as there is a positive result of +5, there is a gain all around.

Cooperative and non-cooperative games can be zero-sum or non-zero-sum games. An example of a cooperative game is Stag Hunt. Two hunters are given a choice of whether to hunt for a rabbit or stag. On their own, a hunter can successfully hunt for a rabbit. However, there is little meat, and thus, little reward to be gained by hunting the rabbit. On the other hand, hunting a stag cannot be accomplished alone. The two hunters need to work together to achieve this feat. There is much benefit to going for the stag, as it has a lot of meat. The Nash equilibrium for this game is cooperation. Both hunters should work with the other to bring down the stag so that individually they can both benefit more. The Stag Hunt typifies a cooperative game, as the circumstances allow for the two hunters to work together. Furthermore, it is also in the hunters' best interests to do so because they can achieve Nash equilibrium.

The Stag Hunt can resemble business activities such as production or running a household. In the case of the latter, if family members each do a chore such as cooking, doing dishes, and sweeping, the family members overall benefit from each of the chores being done. They get to eat a decent meal and have a clean kitchen. Furthermore, they do not need to take responsibility for all the chores. They only need to do their portion.

On the other hand, the prisoner's dilemma—which will be covered extensively in the next chapter—is a non-cooperative game. The circumstances do not allow for the individuals to work together.

Perfect, Imperfect, and Incomplete Information

Perfect information refers to knowledge of the strategies and outcomes available to all the players. Chess is an example of perfect information, as both players have complete knowledge of all the moves available to the other player. Furthermore, it embodies a complete-information game, as the players know the outcomes of the other player. For example, the players have to bring about a checkmate or capitulation from their opponent. Alternatively, poker is a game in which the decisions of players remain hidden. However, the knowledge is complete, as we know what the outcomes or payoffs are of the other player. They win the pot, or they do not.

William Spaniel, who lectures on game theory, explains that it is essential not to confuse perfect and complete knowledge with one another. The same applies to imperfect and incomplete knowledge.

Perfect and imperfect are connected to the possible strategies. Complete and incomplete knowledge relates to the outcomes. Remember, there are three basic features of every game: players, strategies, and outcomes. It is critical in game theory logic to not mistake strategies for outcomes.

Normal Form and Extensive Form

Normal Form and extensive form can also be referred to as sequential and non-sequential games. As explained in the previous chapter, the strategies are affected by time. If the two agents make a decision simultaneously, this follows a normal-form game template. Generally, a matrix is used to represent the possible decisions of both agents. The Nash equilibrium determines which possible strategies produce the best outcomes for both players.

Chicken is an example of a normal-form game. The two players are implementing their strategies at the same time. Chicken is a game in which two drivers are accelerating right at each other. The objective of the game is to demonstrate that one is more courageous than the other.

If Driver A swerves to avoid collision before Driver B, they risk being called a coward and losing face. Conversely, if Driver B swerves first, they also may end up being called a chicken. However, if both drivers fail to swerve, then the outcome involves a collision of the two cars, and the two players may die. The predicament of both drivers may seem like an unrealistic one, but for the purposes of a game, it is a real one.

Chicken is seen as an example of appeasement and conflict strategies for foreign policy. Using mathematics, applying a mixed-strategy approach, the Nash equilibrium for this game is that Driver A will drive on the course for collision one out of 50 times and swerve 49 out of 50 times. The same applies to Driver B. Even with the Nash equilibrium, collision is not always avoided. If the chicken is performed repetitively, a collision will occur. However, Nash equilibrium has produced driving straight on one out of 50 times and swerving 49 out of 50 times as the optimum results.

An extensive form involves turn-taking. Player A takes turns. Player B can implement their strategy based on knowing Player A's strategy. Extensive form is represented as a tree diagram. Each branch or node shows the strategies the player takes. As one player has perfect knowledge of the other's strategies, extensive-form games include perfect-information features.

Chess is an extensive-form game, as there is an invariable number of strategies which each player can implement. Naturally, if the other player succeeds in bringing about a checkmate, the game ends. Yet, the number of moves per game is not constant, like with chicken.

The Utility of Game Theory

As we saw in the introduction when we analyzed the envy-free cake-cutting game, the cake represents resources. Working out the Nash equilibrium gets us to apply mathematics to determine how to go about sharing the resources.

Many of the interactions in game theory present very unique or even strange situations. We have already looked through some of these situations: the traveler's dilemma, and chicken. They do not seem to bear much resemblance to the social interactions we encounter every day. This is not exactly true for the volunteer's dilemma. We are often bombarded with requests for help from friends and acquaintances. These requests may interfere with our lives, causing us to incur losses. Therefore, the volunteer's dilemma explains that sometimes we are in a position where saying "no" is more appropriate, and consequently, we should say "no." We should not help out. The volunteer's dilemma offers individuals valuable information. It indicates that we should not take on tasks which are burdensome or which negatively affect us.

Chicken is another situation which seems greatly removed from reality. However, this is not, in fact, true. In the case of two aggressive countries, where there is potential for war, the chicken scenario provides useful strategies for the leaders of the countries. As mentioned, the Nash equilibrium is that Driver A should continue on the course for collision one out of 50 times and swerve 49 out of 50 times, and Driver B should do the same. While it is true that war will not always be avoided, the Nash equilibrium provides the optimum results. Conflict in most cases will be avoided, and both countries reduce their possibility of losing face. In foreign policy, "losing face" presents much more serious consequences. These nations are seen as weak or incapable of defending themselves.

Thus, even though the games or social interactions presented in game theory seem peculiar, they provide assistance with managing our day-to-day affairs. Like with all games, they help us acquire skills so that we can live our lives to the best of our abilities, but also make decisions which promote the best outcomes for everyone involved.

Summary

Games originated and have evolved in an organic manner in society. It is not by chance that play is present in cultures and nations throughout the globe. This worldwide phenomenon indicates that it is psychologically meaningful to play games or to create social interactions, with imposed rules, to try to either compete or cooperate to attain specific outcomes. The study of games tries to understand why games are important to people and what strategies produce the best outcomes for oneself and everyone else. The best strategies may be selfish sometimes, or they may involve appearing cowardly. However, the Nash equilibrium also attempts to achieve the best outcomes for all the parties involved.

Chapter 3: Prisoner's Dilemma

Introduction

In each of the next chapters, we are going to analyze some of the most popular games or social interactions in game theory. We will describe the situation, look at how it unfolds, and then finally look at the Nash equilibrium for each situation.

Then, we will look at how each of these games parallel some events in real life and how understanding these games can help us to make better choices in both smaller and larger social interactions.

Keep in mind that there are some games which you do not choose to play. In other words, there are some situations you will find yourself in which you do not want to be in. The prisoner's dilemma is one of those.

The Prisoner's Dilemma

The prisoner's dilemma is one of the cornerstones of game theory. If you begin to study this topic, this is one of the first principles that you will encounter. In 1950, two mathematicians, Merrill M. Flood and Melvin Dresher, came up with the prisoner's dilemma principle while working at an American think tank.

Two members of a gang rob a bank. They have been arrested and are now being held in isolated interrogation rooms. There are no witnesses to the crime, so the police rely completely on a confession to legally charge the two robbers. The police officials aim to persuade one of the prisoners to confess, and thus, betray his partner. Therefore, both Prisoner A and B have to choose whether to confess to the crime and cooperate with the police or to remain silent and work with their fellow robber. As the two prisoners are being held in separate interrogation rooms, they have no idea what their counterpart is doing. In this way, the prisoner's dilemma is a game of imperfect knowledge, as the strategies each prisoner takes is hidden while the other prisoner deliberates.

On top of this, Prisoner A and B do not know each other very well. They worked in cahoots in the robbery for financial gain, but they are not related or good friends. Neither has good reason to trust the other. Furthermore, both know that the authorities need to secure a confession from one of them.

If Prisoner A confesses, betrays Prisoner B, and works with the prosecution of the case, they will walk free on the condition that the other does not confess. If this occurs, Prisoner B will receive 10 years. The same applies to Prisoner B. If Prisoner B confesses, betrays Prisoner A, and works with the police, they walk free on the condition that Prisoner A stays silent. Then, Prisoner A will get ten years. If both the prisoners stay silent, they will go to prison for just two years. Finally, if both confess, they will get the maximum punishment of five years.

As the two prisoners are completely ignorant of what is happening in the other interrogation room, they do not know what deal their fellow robber is making with the police. If they do not admit their guilt to the police, and the other prisoner does, they will get the maximum sentence.

The Nash equilibrium for the Prisoner's Dilemma is for both prisoners to confess and betray the other prisoner. The incentives of staying quiet do not outweigh those of confessing. The optimum results are for both prisoners to admit their guilt.

Though staying silent for both the robbers would result in the least number of years in prison all round, it is too risky for either of the individuals to stay silent. They may end up receiving the maximum sentence. They do not know if the other is cooperating with the police in the hope of walking free.

The Nash equilibrium for the prisoner's dilemma is that it is best to act in one's self-interest at the expense of the collective or the other person, especially in such conditions in which you cannot cooperate.

The Utility of the Prisoner's Dilemma

It is not always in our best interest to act selflessly or to the benefit of others. The prisoner's dilemma is the perfect illustration that working in one's own self-interest is actually the least risky option. What is interesting about this principle is that the Nash equilibrium is made considering the worst possible outcomes. If you confess and betray the other prisoner, you walk free. You are lucky. This is not what the Nash equilibrium is seeking to achieve. It is trying to avoid you spending the maximum ten years in jail—the worst possible outcome. There are some situations or contexts we encounter in our everyday life that, if we decide, we should follow the prisoner's dilemma principle.

Ecological Crisis

The prisoner's dilemma applies to the current ecological crisis. Though as controversial as it is, the best strategy in the prisoner's dilemma, according to the Nash equilibrium, is to act in one's own interest.

Country A is given the option of shutting down all industries and powerplants which create CO_2 emissions. If they do so, they ensure greater longevity of the planet. However, they also weaken their economies and decrease their production. For example, closing power stations results in less power being generated. If Country B does the same, then both countries are weakened. And the same applies in the opposite sense.

Country A does not know if Country B will in fact go through with the policy changes. They may choose to use the other country's choice to close production as an opportunity to gain economic advantage. Country A is not fully aware of whether Country B is really adopting environmentally friendly policies. Thus, country A is at risk of falling behind, while Country B is at risk of getting ahead.

In international relations, this scenario is playing out. Russia seeks to benefit from global warming. The Arctic Ocean, which is closed off to Russia most of the year, will melt, providing them access to better trading opportunities. Furthermore, developing countries such as Pakistan, Kenya, and Sri Lanka wish to advance infrastructure to help their people by increasing their production of energy. Pakistan, for instance, has a deal with China— the China-Pakistan Economic Corridor—to hydroelectric and coal power stations as a means of providing power to their people and reducing the rate of poverty. Even with the looming environmental crisis, if they do not act in their own interest, they will continue to remain developing countries and fall behind on an international scale. These governments also aim to keep their populations satisfied, as, if they fail to, they themselves could be voted out and replaced with a party who will invest in the country's development.

It is a contentious issue. Some thinkers like Yuval Noah Harari state that the environmental crisis is an issue of such immense magnitude that it transcends national interests, thus superseding game theory logic. Nevertheless, the prisoner's dilemma seems to be playing out in the global response to the ecological crisis.

Economics and Business

A prisoner's dilemma situation can transpire in business. There are two companies who are competing with another for dominance in the market. For the point of this example, they are Adidas and Nike.

Adidas wants to increase their market share by dropping their prices. If Nike retains their original prices, they will lose their share of the market, as there will be more sales of Adidas items. As a result, Nike has no option but to decrease their prices to stay competitive.

The Nash equilibrium for competing companies is to drop their prices to retain their market share or to make a grab for even greater market domination. Interestingly, Nash equilibrium produces the optimum outcomes for Adidas and Nike and for the consumers. Both companies have to lower their prices to remain competitive. With lower prices, customers spend less. People always aim to spend less, so they will go for the more affordable option.

There are limitations to the prisoner's dilemma in business. As cooperation is not a feature of the prisoner's dilemma, both companies cannot collude. While it is true that rival companies do not often communicate—to maintain a competitive edge over the other—there have been instances of price-fixing.

> Price fixing is an agreement between participants on the same side in a market to buy or sell a product, service, or commodity only at a fixed price, or maintain the market conditions such that the price is maintained at a given level by controlling supply and demand.

Price-fixing allows for collaboration between the businesses. This will be detrimental to the customers. However, as collaboration is possible, it does not epitomize the prisoner's dilemma. Yet, even if Nike and Adidas agree to price-fixing, they do not know if their competitor will keep their word. They may become greedy and still attempt to get further domination over the clothing market.

Politics

The prisoner's dilemma also can be handy when making decisions in political scenarios. There are two political parties competing for votes. Both parties are aware that they need to cut government spending. As the government does not produce any wealth, it relies on taxpayers to pay the debt. However, if the debt is too high, the party will lose votes.

If Party A takes steps to mitigate the growing debt, they will gain more popularity. On the other hand, if Party B is proactive and responds to the national debt issue, they will get more votes. Furthermore, if neither of the two do anything to reduce the deficit, they will both lose voters. Another party, Party C, will appear and threaten the political dominance of Party A and Party B.

Thus, the prisoner's dilemma reveals that it is best for both parties to do what is in their own best interests—to try to gain popularity with the voters. They need to appeal to the voters by reducing government spending and the national debt.

Summary

When it comes to business strategies, responding to the environmental crisis, or a political party's actions to decrease national debt, the Nash equilibrium of the prisoner's dilemma seems to hold true. In situations where businesses, countries, and political parties cannot cooperate or cannot be sure of what strategies their counterparts may be following, and if there is a risk that those methods will involve great loss to oneself, it is wise to act in a way that is most advantageous to oneself.

If the Nash equilibrium is applicable even in much larger social interactions, it is useful even with the smallest of social interactions. Thus, if you are in a situation where you do not know the strategy of the other player. You may also not know the other individual very well. If the other player's strategy poses great risks, and you are given the opportunity to take action to minimize the losses, then you should do so. People already do this all the time. They create leasing agreements, do not let strangers into their houses, and pay for insurance for peace of mind.

Chapter 4: The Shapley Value

History

In 1951, Lloyd Shapley, an American mathematician, worked on his thesis about how to solve the problem of distribution in a cooperative social instance. The Shapley value is now considered one of the cornerstones of game theory as it analyzes cooperative games in more detail.

In short, the Shapley value ensures that each individual earns as much or more from a collaborative activity than from working independently. In other words, from an instance of cooperation, there must be an incentive to collaborate or an equal or greater return of investment to working with others than from working alone.

The Shapley Value

There are two workers. Worker A can bake ten cakes in one hour. Worker B can bake twenty cakes in an hour. If they decide to work together, they can produce 40 cakes per hour. Alternatively, if they worked alone, there would only be a total of 30 cakes. In this instance, worker A takes on part of the duties such as preparing the ingredients and mixing the batter, while worker B takes on their share of the responsibilities, pouring the cakes into the baking tins and adding the icing. There is much incentive for the two workers to collaborate with baking cakes, as there will be a greater rate of productivity. There will be more cakes.

The workers then decide to sell the cakes. Each cake is sold for $10. The total earnings are $400. In line with the Shapley value, each individual should earn according to their contributions. They will not split the money into two, as Worker B will receive $200 whether they work on their own or in a team. Worker A will receive $200 from the collaboration, but on their own they would have only baked 10 cakes and would have only made $100.

To first work out how much each worker should be paid from the situation, we need to determine each worker's marginal contribution. Marginal contribution is "A value of the group with the player as a member minus the value of the group without the player minus the value created by the player working alone". If we know what value Worker A or B contributes to the process, then we can begin to know how to distribute the gains.

In the above situation, Worker A on their own can bake 10 cakes. We subtract their marginal contribution from the total, which is 40. After subtracting, you get the final amount of 30. For Worker B, they can bake 20 cakes. If you subtract their contribution from the final total, you get an amount of 20. To work out the contribution of both workers to the process, you should subtract Worker A's total amount from Worker B's. Worker A will get a difference of 10. Then, you need to determine the average between the two total amounts. Worker A will make $100 from their 10 cakes, and Worker B $200 from their 20 cakes. If you add the two, and calculate the average, Worker A gets $150. Worker B gets $250. If the Shapley value is added to this process, there is an incentive for both workers to cooperate. Worker A gets an increase of $50, and Worker B also does.

There are two accompaniments to the Shapley value. If two workers or two parties bring the same things to a social interaction, their outcomes or gains should be exactly the same.

The next one is that "Dummy Players have zero value". If someone does not contribute to the overall interaction, they should not reap any gains.

The Utility of the Shapley Value

The Shapley value can be immensely useful in managing social interactions. The above example of Worker A and B can be applied to everyday situations which involve economics and distribution. The situation of Worker A and B works with tangible products such as the creation of cakes (goods) which produce tangible earnings. When we go into more abstract situations, like how much to pay workers who have differing roles or worker benefits, it is more difficult to apply this principle. We will also analyze how the two accompaniments of the Shapley value play out in daily interactions.

Paying the Bill

This is probably one that has been universally applied. It has been derived from Shapley value principles. If two people go to a restaurant, order exactly the same things, then they should split the bill. If a group of people sits down for a meal, but one person does not drink or eat anything, then they do not need to contribute anything to the bill. Finally, if two people eat out, according to the marginal contribution, they should pay relative to what they ordered. Person A orders food and drinks totaling 60% of the bill, they should make a payment of 60% towards the bill. Following this logic, in a group where everyone orders something different, splitting the bill is actually not in line with the Shapley value.

Naturally, there are reasons for this deviation. Some people are good friends and do not mind contributing the same as their friends, even if their meals were less expensive. On dates, it is traditional for one person to pay to show courtesy or even romantic interest. Good friends or relatives may wish to take the other person out for a meal as a treat and a sign of love. Nonetheless, in situations where the individuals do not know each other as well and they have not developed trust or affection among each other, then the Shapley value can be immensely useful for working out how to pay the bill.

Employee Wages

As workers have different skills, and it is hard to judge the exact monetary value of an employee, it is difficult to work out how much each employee should earn. It is not sufficient to pay the workers equally if their marginal contribution is different. Like we saw in the example of Worker A and B, If Worker B earns $200 from their 20 cakes on their own, there is no incentive for them to collaborate with Worker A. Certainly Worker A will be enthusiastic about the cooperation, but if Worker B decides not to contribute to the cake-baking process, then Worker A will go back to making 10 cakes and earning $100.

Shapley's logic of marginal contribution has been adopted by enterprises around the world. For example, engineers, developers, and project managers who manage or create more successful product lines can earn according to the profit of their product lines. If Project Manager A's app generates 24% of the company's total revenue, and Project Manager B's app generates

27%, then their earnings can be worked out accordingly. Their remuneration could increase if they bring in a higher percentage of the revenue, thereby encouraging them to raise sales.

In line with the Shapley value, those who contribute equally to production should receive equal remuneration. Thus, if two employees have the same skills, they should get the same salary. This seems fair practice, but sometimes, it does not play out. For example, some employees have done market research and know they negotiate a higher package, while others do not bargain so hard. There are also some industries where it is difficult to work out if the skills are exactly the same.

In a school, there are two teachers. They both teach six classes. Teacher A is an art teacher and has six classes per day with all the grades to fill their schedule. Teacher B is a first-language teacher who has six classes per day, but with only one grade. Students need to take the language class to meet the national education's requirements for college acceptance. Art is only necessary for those who want to study fine art and art-related subjects at a tertiary level. The demand for both is different. Furthermore, by teaching the first-language subject, Teacher B makes the school fulfil an educational requirement. Art may not be a necessary subject, but it can attract students who have a specific inclination for that subject. The Shapley value would, in this instance, provide a useful basis for calculating what each teacher should earn. Yet, to do so requires thorough understanding of how great a contribution both teachers make to the total revenue of the school.

International Relations

Using the Shapley value for international relations can provoke a controversial response. The USA is the biggest contributor to organizations such as the IMF and NATO. The IMF is a fund where one of its functions is to invest in developing countries. However, as the US is the major financer, in line with the Shapley value, it should reap the most rewards from the fund. Therefore, it should use the IMF to finance projects that bring them profit or offer them some value.

Moreover, with NATO, the USA subsidizes about 70% of the organization. Next is the UK. However, their financial input into the organization is about a tenth less than the US. While some may argue that the USA should have more of a role in deciding the direction of NATO, others say that it should be worked out according to the vote. Every country has one vote. The strategy that receives the most votes from countries should be followed out. If the second option—the more democratic one—is implemented, it has the potential of isolating the major contributor. As we saw with the original example of Worker A and B's cake business, if worker A does not make a profit or if they incur a loss, then they will decline the offer of collaboration.

The resolution of the Shapley value would be to assess the marginal contribution. The proportion of the parties should reap benefits according to the portion of their input. That being said, it does remain a contested solution when it comes to international relations.

Conclusion

The Shapley value has three main principles: marginal contribution, dummy players receive zero for zero contribution, and equal contribution results in equal reward. This principle is applied in scenarios of cooperation. There are various levels of cooperation. For example, deciding on how individuals should pay for a bill, how employee packages should be calculated, and how much say a financer or stakeholder should have in an organization. The main purpose of the Shapley value is to promote fairness and to encourage cooperation. Like with the cake business, cooperation often results in greater production. To reward the greater production, the two individuals need to be rewarded according to what they bring to the process or organization so that collaboration will occur.

Chapter 5: Battle of the Sexes

History

Battle of the Sexes (BoS) is another popular social interaction in game theory. American Mathematicians R. Duncan Luce and Howard Raiffa were known to have first analyzed this game in 1957 in *Games and Decisions: Introduction and Critical Survey*.

As Serrano and Feldman explain in their article, often there is comparison between the prisoner's dilemma and BoS. The difference between the prisoner's dilemma and other game theory situations is that there is an easy solution, or Nash equilibrium. There is not such an easy solution to BoS.

Battle of the Sexes

It should be mentioned that this interaction has experienced some criticism for promoting traditional roles. First, it was developed in 1957, so it is a little conventional. It can be adapted so that "girlfriend" and "boyfriend" can be substituted for "Player 1" and "Player 2." Nevertheless, I will retain the traditional game theory approach in this book.

There are two people who are dating, Girlfriend and Boyfriend. It is a Wednesday night, and they have planned to spend the evening together. However, as they arranged the date some time in advance, they cannot remember if they decided to go to the opera or a football match. As this game existed before cell phones, they cannot phone each other. Girlfriend prefers to go to the opera, while Boyfriend prefers to go to the game. Most importantly, both would actually like to spend the night in the company of the other rather than to spend it alone.

Therefore, for Girlfriend, the most desirable result would be to spend the night watching the opera and in Boyfriend's company. Though she loves the Opera, she does not want to be without Boyfriend's company. On the other hand, though she hates football, if she went to the football game and met Boyfriend, then it would not be such a bad night overall. The worst possible outcome for Girlfriend is going to the football game and having to watch the game alone.

When represented in a matrix, watching the opera and having Boyfriend's company would score 3, watching the opera on her own would be 0, watching the game with Boyfriend's company would be 2, and going to the football game and not having Boyfriend's company would be 0. The same applies to Boyfriend, although his preferred activity is the football match. So, if he spends time with Girlfriend and watches the match, his positive outcome would be 3, going to the football game on his own 0, spending time with Girlfriend and watching the opera would be 2, and going to the opera on his own would be 0.

Thus, in this social interaction, there is much need for the individuals to comply and agree to the other's preferences. The problem is they cannot communicate. BoS is a simultaneous, cooperative game where there is imperfect information. Girlfriend and Boyfriend do not know the other's strategies. They also have to make the decision at the same time.

Nash Equilibrium

What is interesting about BoS is that a Nash Equilibrium has not been created for this. Or rather, it is still disputed among game theorists.

To come up with a resolution, mathematicians follow a similar approach as to chicken. They consider what Boyfriend and Girlfriend should do by repeating this scenario many times. Like with chicken, this game also incorporates a mixed strategy. Remember, a pure strategy means that the players follow only one method and stick to it to bring about the best results. However, a mixed strategy entails that the players can employ both methods. As this is repeating the scenario, this is possible.

The proposed Nash equilibrium is that Girlfriend should go to the opera 3 out of 5 times and Boyfriend should go to the football 3 out of 5 times. They would still have some possibility of meeting each other.

This game has also been adapted. As Girlfriend prefers the opera, it will promote a positive outcome of 1, and as Boyfriend prefers the football game, it will result in a score of 1. Thus, applying the above Nash equilibrium with this adaptation means they would not "burn any money," as the theorists express it.

Another resolution which has been suggested is that the players flip a coin. If we go by the adapted version in which Girlfriend going to the opera and spending the night alone equals a 1 result, then this introduction of chance makes more sense. If the coin lands on heads, then Girlfriend and Boyfriend must go to the Opera. If the coin lands on tails, then it is the football game. However, the mathematics for flipping a coin results in even lower positive outcomes.

The Utility of BoS

BoS presents quite a unique social interaction. Therefore, there are few social interactions which it resembles.

Tandon Pankaj, in an article on game theory, demonstrates a business situation where BoS is applicable. There are two companies, Kia and Hyundai, who need to reach agreement on two compliance standards. The most desirable result will be if they can develop or agree on the same compliance standards. However, neither is willing to adopt the other's standards. Thus, they reach an impasse.

Applying the Nash equilibrium, which entails following one's preferred opinion of compliance standards, is actually one that is demonstrated in business. Many companies do not develop mutually accepted business standards.

> Here experience seems to suggest that the impasse tends to win. IBM and Apple were unable to agree on a common operating system, Sony and the VHS consortium were unable to settle on a standard format for videotape, and in recent years manufacturers of mobile phone handsets have failed to agree on a common system either.

There is some relevance of BoS in some labor issues. Say an employer and a labor union are working together to create a new employee contract. Naturally, they have a conflict of interests. The employer wishes to keep costs low, so they will try to keep the employee wages as low as possible. However, the union aims to raise the wages as high as possible. Neither of the two really wish to follow their counterpart's strategy—like Girlfriend does not wish to watch football games—but the two have to come to an agreement to create the contract. Like with BoS, it is in both the employer and union's interests to come to an agreement.

The Nash equilibrium suggests that the employer and union three out of five times follow their pure strategy—the employer implementing lower wages and the union asking for higher wages. However, in this case, the results produce disappointing outcomes. The two may never agree to the terms of the contract, and the contract may not be created.

Conclusion

What is interesting about BoS is that it shows that even the Nash equilibrium for a social interaction can lead to disappointing outcomes. In fact, even though the Nash equilibrium states that one should apply one's pure strategy or go with their preferred option three times out of five, it has been concluded by many theorists that the results would not actually produce the ideal outcomes.

With the game's adaptations, more positive outcomes can be achieved, as Girlfriend and Boyfriend would enjoy their own preferred form of entertainment more than watching the alternative entertainment option. When it comes to business-compliance standards and employer–union negotiations, tangible results indicate that the Nash equilibrium fails to produce outcomes which are favorable overall.

Chapter 6: The Centipede Game

History

In the last chapter, we analyzed a social interaction in which the Nash equilibrium for the social interaction was disputed among thinkers. In this chapter, we will examine another popular game in game theory that has brought about some interesting developments.

In 1981, Robert W. Rosenthal—an American political economist—devised the centipede game. The centipede game is an example of an extensive, non-zero-sum game where both players have complete information.

The Centipede Game

There are two players, Player A and Player B. Each player takes a turn, starting with Player A. During the first turn, Player A receives $0, and Player B receives $0. Player A must then decide whether to continue with the game and allow Player B to have their turn. If this is the case, Player B gets their turn, and they will receive $3, while Player A will get $1. On the third turn, player A will receive $4 and player B $2. Once again, if Player B has their turn again, on the fourth turn, Player B will receive $5 and Player A $3. The game will proceed in a similar manner. $2 will be added during each turn to either player.

Player A and B will alternate between deciding whether to continue the game or to stop the game to take the cash. The player who chooses to stop the game at any given time will always end up with $2 more than the other. However, there are 100 turns in total. The name centipede game refers to the 100 turns as a centipede has 100 legs.

After 100 turns have elapsed, both players can walk away with $100 each. Therefore, there is much incentive for both players to continue playing right till the end so that they can both walk away with an all-around positive outcome. However, on the 99th turn, Player A has the option of receiving $101 and Player B $99.

Subgame Perfect Equilibrium

For the centipede game, the Nash equilibrium for the game is referred to as a subgame perfect equilibrium. A subgame perfect equilibrium involves using back induction. Causing the optimum choice to be made means that only one move takes place in the game. Thus, an extensive game such as the centipede game is always cut down to one step. Let us look at how this applies to the centipede game.

The most Player A has to gain is on the 99th turn. They will earn $101 and Player B $99. Player A thinks that this is most reasonable, as they will get $101 and also considers that Player B receiving $99 is also advantageous for Player B. Therefore, they will opt to end the game on the 99th turn. However, following this logic, Player B on turn 98 understands that Player A will gain more in the turn. On turn 98, they will get $100 and Player A will receive $98. To them it also seems a fair trade. Thus, they will end the game on turn 98. This is called backward induction. The two players know how the game will play out. Based on how it will play out, they calculate backwards and realize that, at every possible turn, it is better for them to end at any given time, as they always serve to earn money. This is the case with every single round except the first round. On the first turn, Player A and Player B will earn $0.

While it sounds absurd, the subgame perfect equilibrium or Nash equilibrium for the game is actually for Player A to end on the first turn. Even if they walk away with nothing, it is better for them to earn nothing than for player B to go on to turn two, where they earn $2 and Player A receives nothing.

Discussion

Like with BoS, there has been some discussion among researchers and mathematicians about the subgame perfect equilibrium for the centipede game. When experiments of the centipede game were conducted, participants played the game for much longer than the first round. Interestingly, it was when experiments were performed with economists and chess players where the shortest spans of the game were experienced.

Researchers also noted that instances where one of the players was not able to understand the sequence of the game or perform backward induction thinking, they may be inclined to continue playing to see the game through. The other player who comprehends the dynamics of the game better will use this to their advantage. Nevertheless, in such cases, there is a lower likelihood of achieving the subgame perfect equilibrium if the two players are not able to understand the game's dynamics and employ backward induction.

Another topic of discussion that was brought up was the concepts of altruism or reciprocity. If Player A and Player B were related or close friends, they would work together to make sure that they both got to the final round, where both earn $100. As researchers explained, this is an instance of altruism. Working together to make sure to bring about a mutual benefit results in even greater outcomes. The two friends or relatives have worked together to help each other, and therefore, the feelings of trust and reciprocity are enhanced.

Thus, in this case, the centipede game would be an instance of cooperation instead of competition. Like with game theory resolutions, the subgame perfect equilibrium has drawn some criticism, as it seems to propose self-interest as opposed to mutual benefit. This may only be when you play someone you do not know. Therefore, as you have not developed any level of trust between each other, you cannot trust that they will not end the game before the end so that they gain more than you.

The Utility of the Centipede Game

The subgame perfect equilibrium for the centipede seems to reflect what happens in actuality. We will look at some examples from real life to show how mathematics holds true.

Pet-Sitting

There are two neighbors who have a cordial relationship. Neighbor A asks Neighbor B to feed their dogs and water their plants during their time on their holiday. Neighbor B may choose to say no to avoid the inconvenience. However, Neighbor B realizes this is an opportunity for them to go on holiday, and they could then call in a favor with Neighbor A. Consequently,

Neighbor B agrees. And so, a relationship of turn-taking arises between the neighbors.

Yet, this may not be the case. Neighbor B may fulfil their promise and feed Neighbor A's dogs and water their plants, but it may just be that if Neighbor B asks the favor in return, Neighbor A will make an excuse, or worse may fail to do their part, and Neighbor B's plants will die and their pets will starve. This may even apply with the initial request. Neighbor B may not bother at all. Furthermore, one of the players will have to defect eventually. Or they may sell their house and move away. Thus, the one neighbor may gain where the other does not or gain more than the other in the long run.

There is an interesting dimension to this example. The above instance is between neighbors. However, what if an acquaintance you did not know very well asked you to take care of their pets while they were away. You barely know them. In this case, it is better to apply the subgame perfect equilibrium, which means that, if someone does not know you well enough, they should be paying you to pet-sit.

Initiating a Potential Romantic Relationship

This is an interesting version of the centipede game. We have all been in those situations when we see someone we like. We look at them hoping to catch their eye, and if we do, we hope they respond indicating some room for initiation. It could be a smile, a wave, or a wink. We may even be braver from the onset. We may smile at them or try to hold eye contact for a little bit longer. From the onset, we are playing a centipede game.

The person right from the beginning can break eye contact, show no inclination for further communication, and snub us. Person

A will feel hurt and vulnerable, but they will know that they do not have to invest any further time or energy. Person B has made their inclinations clear.

As demonstrated in experiments, it was generally only chess players and economists who ended early. This also reflects everyday life. For example, sometimes, we are not interested in another person. Person A smiles, and Person B smiles back. Person B does not want to be rude. Person A takes this as an instance of reciprocation, but it could be misunderstood. Person A sees the relationship as a potential for a romantic union, and Person B just wants to be friendly. Such things can ultimately end up as a situation of unrequited love, or Person A being stuck in the friendzone. These are very commonplace. In fact, it is very likely that more people have suffered from these outcomes than those who have not. To avoid such a fate, it can be useful to apply the subgame perfect equilibrium from the beginning. You quit as early on as possible.

However, what happens if Person A and Person B both do have romantic inclinations. This is commonplace. It is often depicted in Hollywood films where the two players are afraid to make the first move. Naturally, they do not want to embarrass themselves, or even jeopardize the friendship, so they continue playing the centipede game. Once again, we can turn to the findings from the centipede game. If we know the person well, we can keep playing the game with them. Conversely, if we have only met them, we should get to know them better before investing seriously in them. However, such a resolution seems only to create a further catch 22. We must keep playing to get to know them over time.

International Relations

In Steven J. Brams and D. Marc Kilgour's article "A Note on Stabilizing Cooperation in the Centipede Game," they explain that, after the Cuban missile crisis, a centipede game took place between the USA and the USSR. Unlike Cold War interactions prior to the Cuban missile crisis, there had been no established communication between the two countries. These interactions did not model the centipede game, as it was a game of imperfect information. Brams and Kilgour reveal that "the two superpowers established a 'hotline' in 1963 to enable electronic communication that could forestall a future crisis that might escalate to nuclear war".

The hotline allowed the parties to have communication and receive reassurance from the other that they were not going to deploy any nuclear weapons. Like with all the examples, both parties could not be sure that the other was not going to push the button for the nuclear weapon first. The USA could not know that the USSR had pressed the button. They would wonder if they should push their own to bring about a mutually assured destruction. However, if they continue to play the game, they both aim to create a treaty, where peace is realized.

The subgame perfect equilibrium did not correspond in this case. Both countries had played the game to the end, which saw the end of the USSR. Arguably, the collapse of the Soviet Union could just be seen as one player choosing to abandon the centipede game. Or it is possible that, like many critics propose, that there are instances in reality which supersede game theory strategies.

Conclusion

While many game theory situations seem far removed from everyday interactions, the centipede game is one that mirrors both smaller and larger social situations. From taking turns, to watching one's pets, to making the first move in a friendship or relationship, to instances such as the Cold War where a country is unsure of their counterpart's motives, the same pattern emerges. When the game begins, each player can at any time call quits, resulting in the other player incurring some losses. In the case of initiating a romantic relationship, the potential costs are losing face, getting rejected, and creating an instance of unrequited love. In the case of international relationships, the consequences are much graver.

However, the Cold War is a more extreme illustration. In history, there have been examples of nations ending the centipede game and going to war. Nonetheless, there are many individuals who would encourage their leaders to see the game through instead of applying the subgame perfect equilibrium. This may not be such a bad option. If the two nations go to war at an earlier stage, they will not have had time to develop their arms. Therefore, each may experience less devastation by quitting early as the arms race will not have time to evolve.

Conclusion

Since its inception, game theory has taken off. In the contemporary world, modern-day philosophers like Naval Ravikant and Simon Sinek have applied game theory techniques to provide advice to individuals and enterprises. While game theory can be argued as advocating for self-interest as opposed to mutual benefit, there are some instances such as chicken which recommend avoiding conflict 49 times out of 50. Nevertheless, the principle seems to hold true that individuals should go for the option which produces the best outcomes all around. Interestingly enough, if both individuals are acting in their own interests, then the consequences are facilitating optimum results for both. Perhaps this is the rationality behind acting in one's self-interest.

There are critics who propose that game theory is immoral or not acting within a moral capacity. However, I will close the book with John Nash's words. Nash's famous line indicates this is not at all the case: "The Best for the Group comes when everyone in the group does what's best for himself and the group". This is what game theory teaches us to do. This is the art, or rather science, of game theory.

www.ingramcontent.com/pod-product-compliance
Lightning Source LLC
LaVergne TN
LVHW021738060526
838200LV00052B/3344